Intrafascicular Electrodes for the Activation of Peripheral Nerves

Intrafascicular Electrodes for the Activation of Peripheral Nerves

PhD Thesis by

Aritra Kundu

*Center for Sensory-Motor Interaction,
Department of Health Science and Technology,
Aalborg University, Denmark*

River Publishers

Aalborg

ISBN 978-87-92982-63-6 (paperback)
ISBN 978-87-92982-62-9 (e-book)

Published, sold and distributed by:
River Publishers
P.O. Box 1657
Algade 42
9000 Aalborg
Denmark

Tel.: +45369953197
www.riverpublishers.com

Contents

Preface

I wish to thank everybody who has given me the opportunity, assistance and inspiration during my dissertation work in the period of 2009-2013 at the Center for Sensory Motor Interaction, Department of Health Science and Technology, Aalborg University, Denmark. The work was supported by the EU TIME project (CP-FP-INFSO 224012/TIME).

I am deeply indebted to my supervisor, Winnie Jensen for her trust, support, enthusiasm and most importantly patience towards me, until the last moment. It had been a privilege to work under her guidance. Also, I generously thank the co-authors of the scientific papers of this dissertation; Kristian R. Harreby, Ken Yoshida, Tim Boretius, Thomas Stieglitz and Martin Wirenfeldt, for imparting me with knowledge, and interesting and exciting ideas in various aspects whenever I needed.

Particularly, I would also like to convey my special thanks to Ole Sørensen, Torben Madsen, Jens Sørensen, Birgitte Jensen and Mogens Vyberg from department of Pathology, Aarhus University Hospital-Aalborg, for assistance in the animal experiments and preparation of nerves for histological evaluation.

My genuine gratitude to all the past and present colleagues and staff members of the department, who rendered invaluable assistance and advice and made always made me feel comfortable and enjoyable, while working here. My heartfelt thanks and hugs to all my dear friends who are always there in good and bad times.

To my parents and my family members, for providing me with unconditional love and support. They are my rock and got me going through in every phase of my life, thank you so much.

Aalborg, February 2013

"The greatest glory in living lies not in never falling,
but in rising every time we fall."
— Nelson Mandela

vii

Summary

The success of a neural electrode is often measured by its performance in the selective stimulation of or recording from individual fascicles within a peripheral nerve without damaging the nerve. Neural prostheses are limited by the availability of peripheral neural electrodes that can be used to record the user's intention or to provide sensory feedback through functional electrical stimulation for an extended duration. Improvement in future neural prostheses may be achieved through the availability of multichannel, bidirectional neural interfaces, such as the transverse intrafascicular multichannel electrode (TIME). The aim of this thesis was to investigate the biofunctionality, biostability, and biosafety of the TIME in an animal model that bears a close neuroanatomical resemblance to the human. To this end, four research questions were formulated: 1) *Is the pig model neuro-anatomically suitable for the simulation of the actions and reactions of a neural electrode in a large human-sized peripheral nerve?* This question was addressed in study 1 in which a comparative morphology study was performed on the median and ulnar nerves of Danish landrace pigs and Göttingen mini-pigs. Nerve specimens were selected at different levels for analysis. The number of fascicles in each nerve was counted, and the fascicle diameters were measured. *2) How selective is the TIME compared to other intrafascicular electrodes, such as the thin-film longitudinal intrafascicular electrode (tfLIFE)?* Study 2 was performed to address this question. A comparative stimulation selectivity study was performed in Danish Landrace pigs with the TIME and tfLIFE. Electrodes were implanted in the median nerve, and sequential electric stimulation was applied to individual contacts. The compound muscle action potentials of seven forelimb muscles were recorded to quantify muscle recruitment. *3) Is the TIME biostable and biofunctional for long-term implantation?* This question was addressed in study 3 in which the chronic stimulation selectivity of the TIME was evaluated on Göttingen mini-pigs. *4) Does the TIME fulfil the criteria of biocompatibility and chronic stability?* Study 4 was conducted to address this question. Immunohistological analysis was conducted on nerve specimens from chronic experiment pigs that had been implanted with TIMEs in study 3.

Danish Summary

En neural elektrodes succes måles ofte på dens ydelse i den selektive stimulation af eller registreringer fra individuelle fascikler inden for en perifer nerve, uden at den beskadiger nerven. Neurale proteser begrænses af tilgængeligheden af perifere neurale elektroder, der kan anvendes til at registrere brugerens intentioner eller give sensorisk feedback gennem funktionel elektronisk stimulation i en længere periode. Forbedringer i fremtidige neurale proteser kan opnås ved hjælp af tilgængelige multi-kanal, bidirektionale neurale interfaces som f.eks. tværgående, interfaskikulære multi-kanal elektroder (TIME). Formålet med denne afhandling var at undersøge biofunktionalitet, biostabilitet og biosikkerhed for TIME i en dyremodel, som har stor neuroanatomisk lighed med mennesker.

Til dette formål blev der formuleret fire forskningsmæssige spørgsmål: 1) Er grisemodellen neuroanatomisk egnet til simulation af virkningen og reaktionen af en neural elektrode i en stor perifer nerve i human størrelse? Dette spørgsmål blev behandlet i studie 1, som var et komparativt morfologistudie udført i nervus medianus og nervus ulnaris på grise af dansk landrace og Göttingen minigrise. Der blev udvalgt nerveprøver på forskellige niveauer til analyse. Antallet af fascikler i hver nerve blev optalt og fascikel- diameteren blev målt. 2) Hvor selektiv er TIME sammenlignet med andre intrafaskikulære elektroder, som f.eks. en thin-film longitudinal interfaskikulær elektrode (tfLIFE)? Studie 2, som omhandlede denne problemstilling, bestod af et komparativt stimlationsselektivitetsstudie i grise af dansk landrace med TIME og tfLIFE. Elektroderne blev implanteret i nervus medianus og sekventielle elektriske stimulationer blev påført på individuelle kontakter. Der blev målt sammensatte muskelaktionspotentialer på syv muskler på de forreste lemmer til kvantificering af muskelrekruttering. 3) Er TIME biostabil og biofunktionel til langvarig implantation? Dette spørgsmål blev behandlet i studie 3, hvor den kroniske stimulationsselektivitet af TIME blev vurderet på Göttingen minigrise. 4) Opfylder TIME kriterierne til biokompatibilitet og vedvarende stabilitet? Studie 4 blev udført til afklaring af dette spørgsmål. Immunohistologisk analyse blev udført på nerveprøver fra grisene, der blev implanteret med TIME i studie 3.

1

Introduction

1.1 Background

1.1.1 Neuroprosthetic devices

Start writing and don't stop until it is finished. Surgical amputation of a limb is the final option for the removal of irreparably damaged, diseased, or congenitally malformed limbs, where retention of the limb is a threat to the well-being of the individual [1]. The loss of an upper limb has a profound effect on the daily life activity of an amputee, as the hand experiences a more direct extension of the brain than any other part of the body (known as a somatosensory homunculus). There are 10,000 and 75 cases of upper limb amputations per year in the USA and in Denmark, respectively [2,3].

The hand, and particularly the fingers, is used to perform various tasks and, in the case of the hearing impaired, to express needs through the hand and finger gestures of sign languages [4]. The amputees' dependency on other persons can be reduced greatly if motor and sensory functions can be restored through the use of prosthetics. Body-powered and electrically powered upper limb prosthetics are the most commonly used, but their functionality is limited. These types of prostheses are limited to only one movement at a time and have a limited number of degrees of freedom (DoFs) [5].

With the advancement of robotics, advanced prosthetics have grown to incorporate more life-like functions. The *'Cyberhand'*, which has 16 DoFs, provides individual finger and joint movements [6,7]. The arm build under the *'Revolutionizing Prosthesis'* program [8,9] offers finger, wrist, elbow, and shoulder movements and incorporates sensory feedback to the subject such that the subject can 'feel' the movement of the prosthesis. The technology used in these devices has reached a level that is far superior to our actual current ability to utilise all of the devices' available functions. The effectiveness of and clinical interest in these highly advanced devices depends entirely on the availability of 1) sufficient control signals and algorithms that can translate the control signals into control commands and 2) a sufficient number of channels to provide sensory feedback. One solution that would overcome these issues is to directly interface the peripheral nerves with a highly selective neural interface.

1.1.2   Functional electrical stimulation system

The application of functional electrical stimulation (FES) for neurorehabilitation is on the rise. Paralysis can result from a spinal cord injury (SCI), a stroke, or head trauma [10,11]. Surveys have shown that there are between 250,000 and 450,000 SCI patients in the USA, with approximately 12,000 new cases arising every year [12,13]. Paralysis affects neuromuscular function below the region of injury and blocks the transmission of signals between the central nervous system (CNS) and peripheral nervous system (PNS) [11]. Electrical stimulation of paralysed muscles results in motor function [10]. Through FES, motor fibres are stimulated to activate the motor neurons, or sensory fibres are stimulated to activate reflex pathways [14] for the restoration of lost or damaged limb functions in paralysed patients [15]. FES devices are therefore considered a type of neuroprosthetic device (NP) because they compensate for the lost or damaged functions of muscles and organs [15]. Figure 1 shows a typical motor neural prosthesis setup. Between the 1960s and 1980s, various studies demonstrated the ability of FES to assist paraplegics in standing and stepping [16]. For clinical purposes, FES has been used in the treatment of bladder, bowel, and sexual functions; respiration, restoration of hand grasping and releasing; and standing and stepping [17-21]. Other clinical applications of neural stimulation include bladder prostheses, cochlear implants, retinal and visual prostheses, and vagus nerve stimulation for the treatment of epilepsy and depression [22]. Neuroprostheses for hand grasping and gait modulation use FES to artificially generate the orchestrated muscle contractions that are required in the functional movement of paralysed muscles. Generally, muscle- or nerve-based electrodes are used for FES stimulation of motor systems [19].

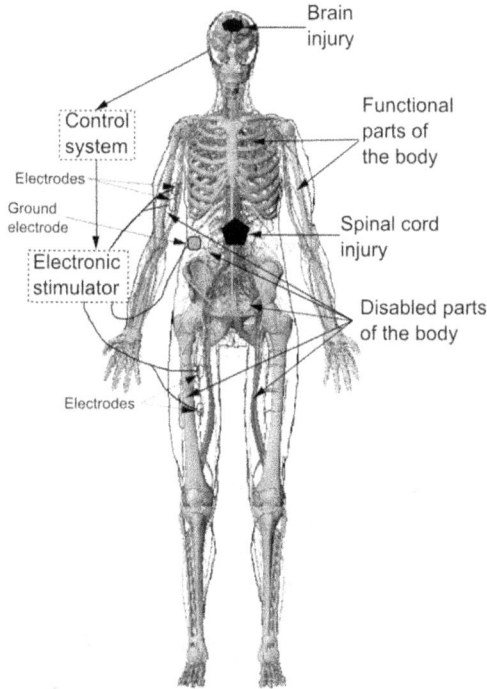

Figure 1: Principles of operation of motor NPs with essential components for control. Adapted from Popovic *et al.* [14].

Neural interfaces play a vital role in determining the success of an advanced neurorehabilitation application. The two most common uses of neural interfaces are in the field of FES and NP devices.
To allow FES systems to perform increasingly dexterous and complex movements, the number of stimulation channels must be increased [23]. Therefore, high-density neural electrode arrays that can selectively stimulate subsets of fascicles, thereby activating only certain muscles or groups of muscles to perform a meaningful functional task, are required [24].
This thesis is based on the evaluation of neural electrodes that can be used for FES neurorehabilitation systems.

1.1.3   Peripheral nerve morphology

The nervous system is an essential and complex biological system of nerves and cells that connects various parts of the body with the brain and spinal cord [25].

Implantation of peripheral neural electrodes requires the understanding of the morphological features of a large human-sized nerve. A nerve is composed of a collection of nerve fascicles bundled by connective tissue referred to as the epineurium. A nerve fascicle is a collection of nerve fibres bundled by connective tissue referred to as the perineurium. A nerve fibre is composed of an axon, myelin sheath, and cellular covering (neurolemma). From the high-magnification hematoxylin and eosin (H&E) stain of the nerve cross section in Figure 2, these structures can be identified.

The epineurium (stained dark pink) can be observed at the top of the image (Figure 2a). Fascicles can be identified within the nerve in the image, each surrounded by a perineurium (stained bright pink). Each fascicle holds hundreds of individual nerve axons (stained blue/purple) (Figure 2b).

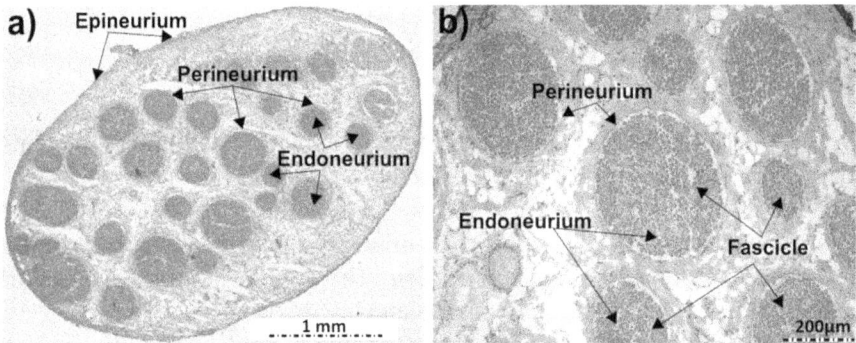

Figure 2: a) High-magnification (20x) nerve sample of a pig median nerve stained with H&E. The epineurium, perineurium, and endoneurium have been marked. All fascicles can be identified. b) Each fascicle holds hundreds of individual nerve axons.

A human peripheral nerve (median and ulnar) consists of an average of 37 fascicles [26]. Therefore, a suitable animal model should be chosen such that the outcome of the experiments performed on the peripheral nerve of the animal model directly reflects the functionality of an electrode implanted in a human.

1.1.4   Neural interface

The NP interface has been a major technological hindrance to the achievement of a successful interface with the nervous system. With the advancement of technology, scientists are eager to find a method to interface and control artificial prostheses with biological signals. The search for such a method has led to the development of the fields of FES and neuroprosthetic devices [7,27-30]. From the above-defined applications of FES and NPs, we can see the importance of a functionally stable, highly selective, biocompatible interface that can selectively activate muscles to form a well-coordinated movement for a long period of time. Selectivity refers to the ability of the electrode to selectively activate or record from individual nerve fibres or a small population of nerve fibres.

Although motor neural prostheses using FES have been established as tools for the neurorehabilitation of SCI and stroke patients, these prostheses are still not popular in clinical use, as there exists an imbalance between the complexity of the technology and its ease of use [23]. Once implanted, these interfaces should be functional for a long time and should be biocompatible with the tissue in which they are implanted [31].

1.1.5   General requirements for an ideal peripheral neural interface

An ideal NP interface demands certain characteristics if it is to become a reliable and robust neural interface for a variety of applications. It should have 1) minimum invasiveness but maximum selectivity, 2) biocompatibility or neurocompatibility [32], and 3) chronic stability, i.e., it should be functional over a long period of time without causing any damage to the surrounding tissue in which it is implanted.

Invasiveness versus selectivity has long been an issue for successful implantable neuroprosthesis. This trade-off requires that the maximum amount of information be acquired from a specific nerve fibre while causing the minimum amount of damage, allowing access to information from sensory afferents, selectively stimulating multiple nerve fibres, and providing control over muscle activation [33]. The trade-off between selectivity, i.e., the ability to activate single nerve fibres, and invasiveness exists for every neural interface [7,30,34,35] (see Figure 3). Thus, an optimisation must be achieved between increasing invasiveness to obtain adequate selectivity and minimising invasiveness to lower the impact to the body, which also influences the histological response of the body and the chronic stability of the interface.

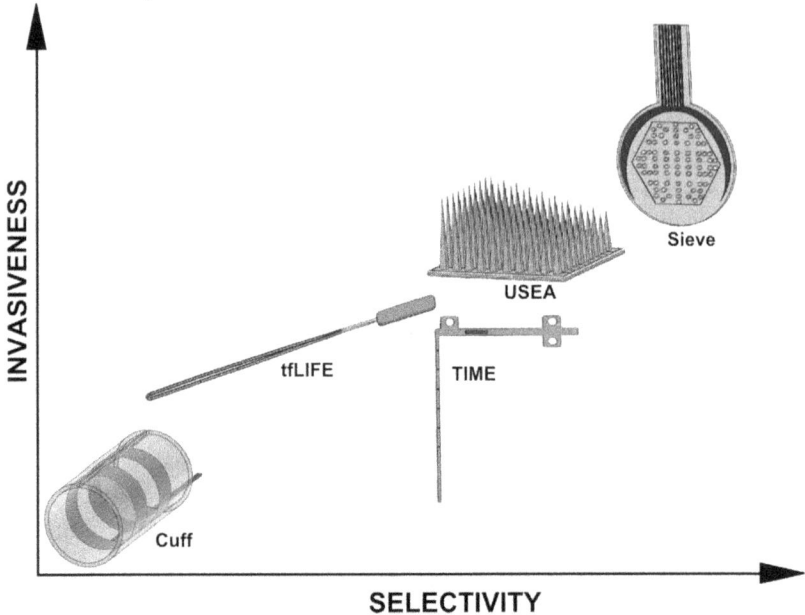

Figure 3: Graph showing invasiveness versus selectivity for different
types of neural electrodes (see text for a detailed discussion)

The concept of biocompatibility is not limited to the non-toxicity of the implantable electrode but also encompasses the physical and chemical surface properties of the electrode and the reaction of the tissue surrounding the implant. Biocompatibility can be subdivided into three categories: a) *biosafety*, i.e., the implant does not attack the host tissue; b) *biofunctionality*, i.e., the implant works smoothly within the tissue; and c) *biostability*, i.e., the implant is immune to attacks from the surrounding tissue and fluids [32].

To achieve chronic stability, a neural interface should have mechanical and geometrical properties that exactly fit the implantation site to reduce the effect of traumatic lesions [32].

Peripheral nerve interfaces can be classified as either a) *extraneural*, i.e., those implanted around the nerve trunk, or b) *intraneural*, i.e., those that penetrate the nerve trunk.

Compared to other electrode types, such as muscular and surface electrodes, neural electrodes require lower stimulus intensity for activation of neural tissuedue to their proximity to the nerve, which simultaneously lowers the potential hazardousness of electrochemical processes and the power consumption of the stimulator systems [7,36]. The implantation of neural electrodes is invasive but requires a small surgical procedure. Because a peripheral nerve electrode is placed away from muscles, this type of electrode has certain advantages over intramuscular electrodes (IMs) for FES. There is no effect of mechanical stress on the neural electrode due to muscle contraction. The peripheral nerve innervates several muscles; a neural electrode is comparable to the implantation of several muscular electrodes. The charge required by an IM for muscle stimulation might be 10 to 100 times higher than that required by a nerve electrode [37].

Neural electrodes are independent of muscle length and electrode placement recruitments [38,39]. Electrodes implanted within a nerve are intraneural electrodes and can directly access the nerve fibres that are needed to stimulate or record. These electrodes typically have a high signal-to-noise ratio (SNR) and increase the probability of stimulating and recording selectively [7,40,41].

From the above two applications (FES and NPs) of neural interfaces and the advantages of using intraneural electrodes, it is possible to prioritise certain characteristics that will determine the efficiency of an electrode. 1) Its ability to interface with the maximum number of fascicles such that stimulating these fascicles can activate additional muscles. 2) The combination of the factors of selectivity, stability, and longevity such that well-coordinated muscle movements can be executed by selective activation of different subsets of nerve fibres over a long period of time. 3) The electrode's ability to cause no or minimal nerve scarring when implanted. The tissue surrounding the electrode should also accept the electrode with minimal resistance and foreign body response.

1.2 State-of-the-art of neural electrodes

This project focuses on the selective electrical stimulation of nerve fibres. To achieve improved selectivity of nerve fibres it is necessary to have an electrode that can only stimulate the desired nerve fibres present inside the nerve with limited invasiveness. Figure 4 shows the hierarchy of the types of neural electrodes used by different groups for their experiments over time.

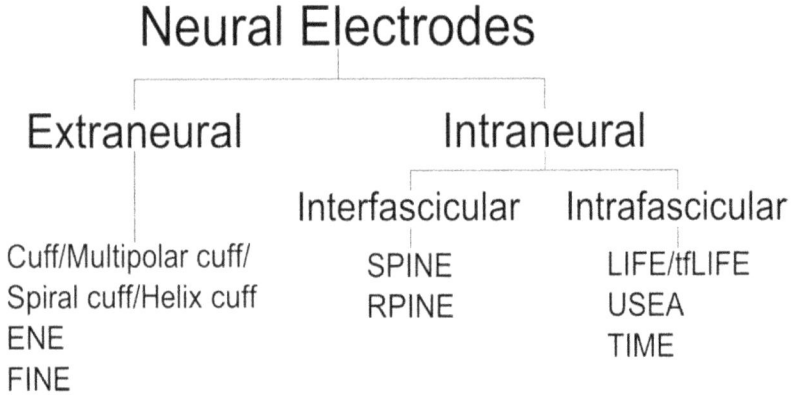

Neural Electrodes

Extraneural   Intraneural

                              Interfascicular   Intrafascicular

Cuff/Multipolar cuff/   SPINE   LIFE/tfLIFE
Spiral cuff/Helix cuff   RPINE   USEA
ENE   TIME
FINE

Figure 4: Hierarchy of neural electrodes

1.2.1   Extraneural electrodes

1.2.1.1   Cuff/Multipolar cuff/Spiral cuff/Helix cuff

Cuff electrodes are examples of extraneural electrodes. They contain two or more active sites embedded in their inner walls. Cuffs are the most frequently investigated neural interfaces in basic and applied research [7]. Cuff electrodes have several advantages over intramuscular surface and epimysial electrodes, including precise positioning of the electrode away from the target muscle, thereby limiting mechanical damage and lead failure during muscle motion. Additionally, the stimulation current is confined to a certain region of the nerve and is significantly lower than that required by other electrodes [7,37]. To take advantage of the current-distance relationship, multipolar cuffs have been designed and implemented. In these cuffs, multiple contacts can be placed around the nerve, and selective stimulation can be applied to the nerve fibres lying close to the active sites without exciting the neighbouring fibres. Acute and chronic experiments in different animal models and human trials have proven that cuffs have the stability and selectivity to activate muscles innervated by fascicles that lie close to the cuff-contacts [31,42-48]. FES through peripheral nerve cuffs has been used therapeutically for the restoration of lower-body functions in paralysed patients [49-51]. The disadvantages of using cuffs are the size of the cuff and the cuff's inability to interact with axons in the centre of the nerve. Compression injuries and the restriction of blood flow at the peripheral nerve can cause neurapraxia if the cuff squeezes the nerve. Loose fitting cuffs cause a reduction in selectivity and require the usage of higher stimulation currents. Although cuffs are used for electroneurograms (ENGs), they can also receive noise from their surroundings, particularly due to electromyographic activity (EMG) [52]. The peripheral nerves in limbs undergo wide ranges of motion and stress [7,33,53]. This motion can exert forces on a cuff, causing nerve-cuff friction that can result in neural damage. Chronic studies with cuffs revealed abnormalities in nerve morphology [54]. The primary disadvantage of the cuff electrode is the determination of a perfect dimension for the electrode such that it fits on the nerve. An irregularly shaped fitting could result in nerve injury and affect the nerve-cuff interface [55].

1.2.1.2   ENE

Epineurium electrodes (ENEs) are sutured to the epineurium of the peripheral nerve. ENEs can achieve controllable, progressive muscle contraction with minimal muscular fatigue during FES. ENEs were used in clinical applications for the treatment of paraplegic patients. Due to proximity to the nerve and reduced electrode dimensions, there was less electrode corrosion and neuronal loss during the use of these electrodes [7,56]. However, ENEs cannot access fascicles deep inside the nerve.

1.2.1.3   FINE

To gain access to the nerve fibres in the centre of a nerve, a flat interface nerve electrode (FINE) was developed [57,58]. FINEs are an alternative version of the traditional cuff electrode [58]. FINEs are oblong in cross section and reshape the nerve when implanted. Acute and chronic experiments on cat and rat animal models and a human model demonstrated that these electrodes could stimulate sciatic and femoral nerves and selectively activate the muscles innervated by those nerves [13,57,59-61]. Although there were no side effects observed from the reshaping of nerves using small amounts of force, large forces applied to nerves can cause neurapraxia. Axon demyelination has been reported at pressures as low as 10 mmHg [62]. During long-term implantation (>28 days), the formation of connective tissues around the nerve and between the nerve-electrode interface can increase the pressure on the nerve, leading to demyelination. FINEs implanted in human femoral nerves have been reported to selectively and independently activate four out of six muscles innervated by the femoral nerve [13]. In the upper limb the peripheral nerves innervate more muscles and have to control complex finger and joint movements. Therefore, FINEs may not be the first choice for implantation because they can activate only a limited number of muscles. A multichannel intrafascicular electrode with several active sites can solve this problem by accessing all of the critical fascicles present in a nerve.

1.2.2   Intraneural electrodes

An intraneural electrode pierces the epineurium of the nerve. Popovic *et al.* have reported on several studies demonstrating that intraneural electrodes can elicit muscle contraction with much lower stimulation currents than those required by extraneural electrodes, such as cuffs [14].

1.2.3   Interfascicular electrodes

1.2.3.1   SPINE / RPINE
Slowly penetrating interfascicular nerve electrodes (SPINEs) and radially penetrating interfascicular nerve electrodes (RPINEs) are interneural electrodes that lie between extraneural and intraneural electrodes. These electrodes penetrate the softer epineurium of the nerve but not the tougher perineurium that protects the fascicles [37,63,64]. Functional neural stimulation (FNS) using these electrodes implanted in the sciatic nerve of a cat resulted in the selective activation of muscles and generated a different pattern of recruitment than extraneural electrodes. The studies performed with interfascicular electrodes were all acute and histologic evaluations showing reorganisation and alteration in nerve cross sections [37], which may involve adverse effects, such as neuropathy, during chronic use of the electrode.

1.2.3.2   Intrafascicular electrodes

The Longitudinal Intra-Fascicular Electrode (LIFE) [65,66], thin-film LIFE (tfLIFE) [67], Transverse Intrafascicular Electrode (TIME) [40,68], and Utah Slant Electrode Array (USEA) [33,53,69] are examples of intrafascicular electrodes.

1.2.3.3   USEA

The USEA is a micromachined silicon structure of needle electrode matrices, which were developed as an interface for the brain and later modified for applications in the peripheral nerve. USEAs have a high active site density and record from the tip of needles, similar to columns that remain in the nerve. Studies using USEAs have shown selective activation of the hind limb muscles of cats when implanted in the sciatic nerve for both acute and chronic studies. The activation current used by USEAs is significantly lower than that used by cuffs, with more graded recruitments of individual muscles [33,53,69]. Although the USEA has a high number of channels, its rigid tips cause continual irritation and even damage to the peripheral nerve during chronic implantation [70]. The USEA requires a pneumatic impulse inserter to implant the electrode in the peripheral nerve, and a containment system must be used to anchor the electrode in position for chronic use. These requirements increase the potential for neuropathy or injury. Histological images after USEA implantation have shown crest and trough formations on the nerve [53].

Investigations of tissue responses to foreign bodies have shown that the design of an implant has an effect on its biocompatibility with the tissue surrounding the implant location. The tissue response is greater when the electrode has a large cross section in addition to sharp features, but the tissue response lessens when the implant has smaller, flexible features [31,71]. Because the USEA has sharp, needle-like, stiff active sites, it may not be suitable as an intraneural electrode.

1.2.3.4   LIFE

LIFEs are a potentially useful neural interface platform for basic electrophysiology and functional electrical stimulation [72] with the ability to activate nerve fibres in different fascicles independently [65] [73]. Studies have been conducted with LIFEs implanted in the peripheral nerves of both animal (rats, cats, rabbits, and mudpuppies) and human models to perform selective stimulation and recording over both short and long durations [65,66,72-77]. Amputees implanted with LIFEs at transradial levels in the residual peripheral nerves (median and ulnar) have demonstrated the sensation of touch proprioception and finger control after electrical stimulation of sensory pathways [78,79]. LIFEs were used to detect volitional motor nerve activity corresponding to missing limb movements [80]. Studies conducted by Dhillon and Horch demonstrated that appropriate, graded, distally referred sensations can be induced through the stimulation of amputee nerve stumps with LIFEs and that these sensations can be used as feedback information on force and limb position [81]. For these experiments, LIFEs were implanted for an average of 23 months in amputee subjects. Histological evaluations of LIFEs implanted in the sciatic nerves of rats for three months have shown signs of local scarring but no nerve damage [82].

These examples demonstrate that LIFEs are biocompatible and capable of selectively activating some peripheral nerve fascicles for long durations.

1.2.3.5   tfLIFE

The tfLIFE initiated the development of a functional multichannel microfabricated LIFE structure [35,67,83]. The aim of this electrode was to increase the number of channels, selectivity, and reproducibility of wire-based LIFEs by microfabricating multichannel structures on patterned and metallised thin polymer substrates. The tfLIFEs are implanted longitudinally along the nerve. They are flexible, have no sharp edges, and are considered less invasive than USEAs.

Several types of acute and chronic stimulation and recording experiments with tfLIFEs have been performed on animal models and in humans. Stimulation experiments on pigs demonstrated the ability of tfLIFEs to activate independent muscle groups (both extensor and flexor muscles) of the pig's forearm by different active site combinations of the electrode and the amplitude of the applied stimulation [84,85]. Neurobiological evaluation of tfLIFEs implanted in the sciatic nerves of rats for three months showed that they were biocompatible [82,86].

Recently, it has been shown that four tfLIFEs implanted in the peripheral nerve (median and ulnar) of an amputee over a period of four weeks produced tactile feedback [87] and control of a robotic hand (with a limited amount of sensory feedback) and significantly reduced phantom limb pain (PLP) in the amputee subject [88].

1.2.3.6   TIME

The TIME is a combination of the tfLIFE concepts and highly selective microneurography approaches [40,89]. In contrast to the tfLIFE, which is planted longitudinally along the nerve, the TIME is designed to be implanted transversally, thereby enhancing the possibility of addressing a much larger subpopulation of fascicles [40,90]. Neural implantation of the TIME results in placing several active sites intraneurally and interfacing several fascicles simultaneously. This scheme is comparable to implanting several tfLIFEs in the nerve [91].

Badia *et al.* studied the stimulation selectivity of TIMEs and compared their results with those of LIFEs and multipolar cuff electrodes. All of these experiments were performed in the sciatic nerve of the rat. The sciatic nerve comprises two distinct fascicles that in turn innervate the gastrocnemius medialis (GM), plantar interossei (PL), and tibialis anterior (TA). The results of the study indicated that the TIME was effective at inducing selectivity at both interfascicular and intrafascicular levels, activating all three muscles via the different active sites of one device. In contrast, the LIFE elicited stimulation in the implanted nerve fascicle and activated only the muscle that was innervated by the branch containing that fascicle [91,92].

2

Research hypothesis and focus statement

The tfLIFE is a functional multichannel microfabricated LIFE structure, as demonstrated by Pellinen and Yoshida *et al.* in 2000 [67]. The microfabrication of multichannel structures on patterned and metallised thin polymer substrates was found to increase the number of channels, selectivity, and reproducibility of wire-based LIFEs. However, in the developmental and testing of this electrode, researchers realised that the longitudinal implantation of tfLIFEs did not result in high selectivity. There was a greater chance of recording activity from the same unit with different active sites placed longitudinally along the nerve fibre. To interface with different nerve fibre population and thereby increase selectivity, it was hypothesized to place the electrodes transversely on the nerve, which led to the development of the TIME. When implanted transversally the TIME should come into contact with a greater number of nerve fibres than the tfLIFE, which is designed to be implanted longitudinally. Figure 5 shows the two different implant strategies of the tfLIFE and TIME in the nerve.

There are additional design differences between the TIME and tfLIFE. The entry point of the TIME, i.e., the "mirror-line," is thinner (100 μm) than that of the tfLIFE (160 μm), making it much easier to penetrate the perineurium of the fascicle. In addition, the increased number of active sites in the TIME electrode facilitates recording from different nerve fibre populations. The number of active sites in the TIME (12 sites) is greater than that of the tfLIFE (eight sites). The maximal charge injection capacity for the TIME is $2.3mC/cm^2$ [93], which is higher than that of the tfLIFE ~4nC/contact.

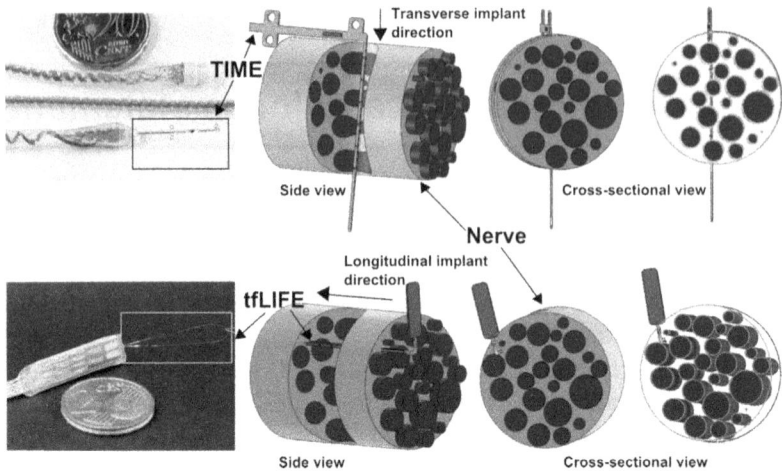

Figure 5: Implantation methods of tfLIFE and TIME structures and cross-sectional views of the TIME and the tfLIFE

Histological and morphological studies on human forelimb peripheral nerves ($Fascicles_{median}$ = 37, $Fascicles_{ulnar}$ = 36) [26] have been performed previously. A literature search revealed the number of fascicles present in the median and ulnar nerves of laboratory animals, such as rats ($Fascicles_{median}$ = 3) [94], cats ($Fascicles_{median}$ = 8) [95], monkeys ($Fascicles_{median}$ = 16±5 , $Fascicles_{ulnar}$ = 2 large + highly variable smaller) [96], guinea pigs ($Fascicles_{median}$ = 4) [97], sheep ($Fascicles_{median}$ = 13 (average)) [98], raccoons ($Fascicles_{median}$ = 4-6) [99], and earthworms ($Fascicles_{median}$ = 1 and 2 lateral giant fibres) [100], but no morphology studies were found in pigs.

Earlier studies on *selectivity* and *biocompatibility* were performed using TIMEs, but the animal model used was the rat [91,101]. TIMEs were implanted in the sciatic nerves of the rats for stimulation. The sciatic nerve of the rat is composed of only two fascicles innervating three major hind limb muscles with a translucent epineurium. Therefore, fascicles can be easily identified, and the active sites of the electrode can be precisely placed inside the fascicle. The implantations in this case were controlled. The results showed that the selectivity of the TIME was significantly higher than that of the tfLIFE and of cuff electrodes and that the minimum threshold activation currents for the TIME and tfLIFE were significantly lower than that of the cuff [91].

Badia *et al.* conducted experiments on the biocompatibility of TIMEs. TIMEs were implanted for two months before histologic evaluation. The results showed an absence of demyelination and a normal axonal count of the fascicles, indicating that TIMEs are biosafe, but no report was made on the microscopic changes to the nerve following TIME implantation [101]. The results from the selectivity study involving rats do not reflect the potential results from the implantation of TIMEs into a multifascicular human peripheral nerve. Therefore, it is important to evaluate TIMEs in a large human-sized nerve animal model before clinical testing.

To investigate the usefulness of the TIME as a neural interface, the following four specific research questions were formulated:

- *Is the pig model neuro-anatomically suitable for the simulation of the actions and reactions of a neural electrode in a large human-sized peripheral nerve?*

A study of pig nerve morphology should be performed to gain insight into the neural structure of porcine nerves and to determine optimal regions for electrode implantation.

- *How selective is the TIME in comparison to other intrafascicular electrodes, such as the tfLIFE?*

It is important to conduct a selectivity study using both TIMEs and tfLIFEs in similar settings to observe their acute functionality.

- *Is the TIME biostable and biofunctional for long-term implantation?*

A chronic stimulation selectivity study with TIMEs should be performed.

- *Does the TIME fulfil the criteria of biocompatibility and chronic stability?*

An immunohistological investigation of a stimulated nerve after chronic experimentation should be performed to determine the biocompatibility of TIMEs.

2.1 Experimental methodologies

All experimental work was performed in animal models. The animal model is presented here.

The experimental procedures were approved by the Animal Experiments Inspectorate under the Danish Ministry of Justice. A total of 21 pigs were used for the four studies. For the acute and morphology studies, 17 Danish landrace pigs were used. For the chronic and biocompatibility studies, four Göttingen mini-pigs were used. The pigs were placed under general anaesthesia for all studies and were laid in a supine position. Access to the median and ulnar nerves in the upper limbs was created through axilla. The heart rate and oxygen saturation of each pig was monitored throughout the experiment.

For histological and biocompatibility analysis, after the nerves were exposed on both sides of the upper limb, the pigs were euthanised using an overdose of sodium pentobarbital. A few minutes after the confirmed death of the animal, the nerves were harvested and specimens were cut at different levels or at the region where the TIME was implanted and prepared for further analysis. After the nerves were harvested from the euthanised animal, the nerves were placed on a wet cloth soaked with 0.9% saline. To maintain the orientation of the nerve specimens for later analysis, the surfaces of the nerve were coloured green and blue for identification of nerves or entry and exit points of the electrodes. If the nerves were not immersed in formalin buffer they were processed for frozen sections. The processing involved placing the nerve specimens on a metal plate, holding them in place with tissue glue (CryoJane; Instrumedics, Inc., St. Louis, MO, USA), and then immediately lowering them into liquid nitrogen to freeze. Both the formalin-immersed and frozen specimens were embedded in paraffin for cutting with a cryostat microtome. The tissue samples were cut to a thickness of 5 μm and later observed under microscopes. Micrographs of the good slides were taken for subsequent analyses.

EMG signals were obtained from a subset of seven left forelimb muscles wherein the left median nerve was stimulated with intraneural electrodes. The muscles were numbered as: Pronator teres (M1), Palmaris longus (M2), Flexor carpi radialis (M3), Flexor digitalis superficialis (M4), Flexor digitorium profundus (M5), Abductor pollicis brevis (M6), and Humeral head of deep digital flexor (M7).

Surgeries for acute and chronic implantations were performed in the left limb, as this setup was more convenient in the laboratory setting for stimulation, recording, and other electronic equipment.

2.2 Summary

2.2.1   Characterisation of large nerve morphology in the pig

The aim of this study was to gain insight into the morphological structure of selected forelimb nerves in the pig. Nerve specimens from six female Danish landrace pigs and Göttingen mini-pigs were analysed. Median and ulnar nerve specimens were harvested from both the upper limbs of the farm pig and the right limb of the mini-pigs. The nerve specimens were cut at three different levels from the proximal through the distal end, i.e., two levels just above the elbow joint and one level below it. A standard H&E stain was used to study the morphological structures. The fascicles were visually identified and counted. The results indicated that farm pigs have a greater number of fascicles than mini-pigs. The fascicle diameter of the mini-pig, although larger than that of the farm pig, was not significantly different. Both farm pigs and mini-pigs have similar median and ulnar nerve diameters as humans. Additionally, the inter-fascicular distance was measured from the two best nerve samples of the pigs.

2.2.2   Characterisation   of   the   stimulation   selectivity   of intrafascicular electrodes

The aim of this acute study was to characterise the stimulation selectivity of TIMEs and tfLIFEs and to investigate the effect of implant angles on selectivity in a large animal model. As demonstrated by Kurstjens *et al.* [85], it is possible to selectively activate the forearm muscles of the pig using tfLIFEs. The primary goal of this part of the study was to implant the intrafascicular electrodes (TIME and tfLIFE) and compare their stimulation selectivity. The peripheral nerve electrodes (TIME or tfLIFE) were implanted in the median nerve approximately 2-3 cm above the elbow joint, corresponding to the approximate implant site chosen for a human implant performed in Rome with tfLIFEs [30,88]. The activation of seven muscles was quantified by measuring EMG signals. To record the EMG signal, bipolar patch electrodes were sutured onto the surfaces of the seven muscles. EMG responses were recorded from seven forearm muscles. The

EMG signals were bandpass filtered (100 Hz - 2 kHz), and the RMS value was calculated in the 2 – 12 ms time interval after stimulation. These values were normalised to the maximum RMS for each muscle to generate recruitment curves. Muscle selectivity was assessed by calculation of the *selectivity index* (SI_M) for each muscle. The selectivity index is defined as the ratio between the recruitment of one muscle and the sum of the recruitments of all muscles, where the contribution of all muscles is equally weighted [46]. We introduced a threshold for recruitment level such that only muscles whose recruitment levels were $\geq$ 30% were evaluated. To quantify the ability of each electrode to selectively activate the group of seven muscles (across all contacts on the electrode), a *device selectivity index* (SI_D) was defined as the average of the SI_M across all muscles. We analysed the *activation current* (I_M) needed to recruit individual muscles at SI_M. A *device current index* for the entire device (I_D) was defined as the average of the IM across all muscles and was used to compare the differences in activation current between the tfLIFE and TIME. The results indicated that both the TIME and tfLIFE required similar I_D values to activate muscles (p = 0.580). Meanwhile, the SI_M and SI_D were found to be higher on average for the TIME than for the tfLIFE (p = 0.019 across all implant angles). The TIME recruited more muscles with higher selectivity than the tfLIFE (a significant difference when comparing the performance of an entire electrode). Meanwhile, SI_M was found to be independent of the implantation angle with respect to fascicle arrangement for the TIME.

2.2.3   Evaluation of the chronic stimulation performance of the TIME in the Göttingen mini-pig

The aim of this study is to investigate the stability and variance of stimulation selectivity of implanted transverse intrafascicular electrodes in chronic use. This study is needed to investigate whether, once implanted, TIMEs remain within the body of the patient for years and work properly. Chronic experiments using TIMEs were performed on Göttingen mini-pigs. Jensen *et al.* [72] and Lago *et al.* [82] demonstrated that the LIFE and tfLIFE, which are similar to the TIME, are biocompatible and demonstrate little functional decline during chronic use. Those experiments were performed through implantation of electrodes in the sciatic nerves of rabbits and rats for three months. In this study, TIMEs were implanted in the right median nerves of four pigs for 33-38 days. During follow-ups, recruitment was assessed via EMGs from five muscles. Formation of a fibrosis layer around the electrode affected the selectivity index and stimulation current. While the device selectivity (SI_D) decreased from 0.25 ± 0.04 in the first follow-up to 0.14 ± 0.05 in the last follow-up, the stimulation current increased from 488 ± 68 μA to 552 ± 48 μA. Throughout the experiments, all electrodes remained in their initial implanted locations. Histological analysis later showed that there were no issues with the biocompatibility of TIMEs and that the healing process of the nerve followed an expected path. The results from this study indicate the need for investigations of the stability of TIMEs over longer durations and possible improvements to the implantation technique so that TIMEs can penetrate the fascicles without bypassing them.

2.2.4   Biocompatibility of subchronically implanted TIMEs

The aim of this study was to investigate biosafety by evaluating the total biocompatibility of the stimulated TIME *in vivo* in Göttingen mini-pigs that were subchronically implanted with TIMEs. Because all implanted medical devices must undergo biocompatibility tests before being introduced clinically, the results from this study will help to establish TIMEs as a potential clinical neural interface. Chronic experiments with TIMEs were performed on Göttingen mini-pigs. Six TIMEs implanted in the right median nerves of four pigs for a period of almost five weeks were analysed for the experiment. Nerve specimens at the implant site were collected for analysis. The thickness of fibrosis around the implant site was found to be 135 ± 56 μm, and the inflammatory cells present were counted at the implant and control regions for comparison and scoring. A modified version of the ISO standard semiquantitative evaluation table was used. This evaluation system depends on the scoring of the number of inflammations within a specified range in a given area per high-powered field (phf). The scoring system was modified according to the high-powered field scaling, and some of the scores for cell responses (such as neovascularisation and fatty infiltrate) were dropped because they were absent when a quantitative analysis of fibrosis was performed. The fields at the implanted region were compared with the fields at the control region. The control region was selected parallel to and approximately 2 mm from the implanted region. Statistical analysis showed that there was a significant difference between the number of inflammatory cells in the implanted region and the control region when all types of cells were considered ($p = 0.008$). The heavy infiltration of fibrocytes at the implanted region caused this significant result (p < 0.001). Fibrocytes are a commonly occurring inflammatory cell during the healing process. When these cells are excluded, there are no statistically significant differences between the implanted and control regions (p = 0.218). This observation indicates that the presence of these inflammatory cells does not have a negative effect on the nerve. To establish the degree of compatibility between the electrodes and surrounding tissue, a modified semiquantitative evaluation was performed, and the results suggested that for all implantations, the TIME was *nonirritant* to its surrounding tissue.

3

Scientific papers

Study 1

Comparison of median and ulnar nerve morphology of Danish landrace pigs and Göttingen mini pigs

Kundu A[1], Harreby KR[1], and Jensen W[1]

[1] Center for Sensory-Motor Interaction, Dept. Health Science and Technology, Aalborg University, Denmark

Corresponding author:
Aritra Kundu
Center for Sensory-Motor Interaction, Department of Health Science and Technology
Aalborg University, Fredrik Bajersvej 7D-3, 9220 Aalborg, Denmark
akundu@hst.aau.dk

Study 2

Comparison of muscle recruitment selectivity during stimulation of the longitudinal or the transverse intrafascicular electrodes (tfLIFE and TIME) in the large nerve animal model

Aritra Kundu[1], Kristian R Harreby[1], Ken Yoshida [1,2] , Tim Boretius[3] , Thomas Stieglitz[3] and Winnie Jensen[1]

[1]Center for Sensory-Motor Interaction, Department of Health Science and Technology, Aalborg University, Aalborg, Denmark; [2]Biomedical Engineering Department, Indiana University-Purdue University Indianapolis, Indianapolis, USA; [3] Department of Microsystems Engineering (IMTEK), University of Freiburg, Germany

Corresponding author:
Aritra Kundu
Center for Sensory-Motor Interaction, Department of Health Science and Technology
Aalborg University, Fredrik Bajersvej 7D-3, 9220 Aalborg, Denmark
akundu@hst.aau.dk

Study 3

Evaluation of the chronic stimulation performance of the Transverse, Intrafascicular, Multi-channel Electrode (TIME) in the median nerve of the Göttingen mini-pig

Kristian R. Harreby[1], Aritra Kundu[1], Ken Yoshida[2], Tim Boretius[3], Thomas Stieglitz[3], and Winnie Jensen[1]

[1]*Dept. Health Science and Technology, Aalborg University, Denmark.,* [2] *Department of Biomedical Engineering, Indiana University - Purdue University Indianapolis ,USA.,*[3]*IMTEK-Department of Microsystems Engineering, University of Freiburg, Germany*

Corresponding author:
Kristian Rauhe Harreby Center for Sensory-Motor Interaction, Department of Health Science and Technology
Aalborg University, Fredrik Bajersvej 7D-3, 9220 Aalborg, Denmark
krauhe@hst.aau.dk

Study 4

Assessment of the biocompatibility of the transverse intrafascicular multi-channel electrode (TIME) following sub-chronic implantation in the median nerve of the Göttingen mini-pig

Aritra Kundu[1], Martin Wirenfeldt[2], Kristian R Harreby[1] and Winnie Jensen[1]

[1]Center for Sensory-Motor Interaction, Department of Health Science and Technology, Aalborg University, Aalborg, Denmark; [2]Institute of Pathology, Aalborg Hospital, Aarhus University Hospital, Aalborg, Denmark

Corresponding author:
Aritra Kundu
Center for Sensory-Motor Interaction, Department of Health Science and Technology
Aalborg University, Fredrik Bajersvej 7D-3, 9220 Aalborg, Denmark
akundu@hst.aau.dk

4

Discussion and Conclusion

The aim of this thesis was to evaluate the performance and safety of a new generation of intrafascicular electrode.
The following research questions were addressed:

- *Is the pig model neuro-anatomically suitable for the simulation of the actions and reactions of a neural electrode in a large human-sized peripheral nerve?*
- *How selective is the TIME compared to other intrafascicular electrodes, such as the tfLIFE?*
- *Is the TIME biostable and biofunctional during long-term implantation?*
- *Does the TIME fulfil the criteria of biocompatibility and chronic stability?*

Four studies were performed to investigate these questions. The outcomes and shortcomings of these studies have been discussed here, and a more detailed discussion of the studies can be found in the relevant papers.

4.1 Methodological considerations

4.1.1 Choice of animal model

Riso *et al.* stated that in the pig model, the peripheral nerves (median, ulnar, and radial) of the upper limb are geometrically and functionally similar to human nerves [102]. As human peripheral nerves are multifascicled (37 fascicles) [26] and innervate a large number of forearm and hand muscles, the evaluation of TIMEs for stimulation selectivity in the pig will likely reflect the performance of the TIME in humans.
The forelimb of the pig has three primary nerves: the median, ulnar, and radial. These nerves branch out at different points to innervate different

muscles. In this thesis, greater attention was focused on the median and ulnar nerves because the radial nerve lies deep within the limb and is difficult to access when surgery is performed through axilla. The peripheral nerve morphology study was conducted on both median and ulnar nerves because these nerves innervate different muscles with various functions, such as flexion, abduction, and pronation. They also innervate the fingers of the hand. All other studies were conducted on the median nerve only due to time constraints and limitations to the number of amplifier channels available during the experiments. Implantation and stimulation of TIMEs in median and ulnar nerves requires placing EMG electrodes on a subset of muscles innervated by these nerves. When TIMEs were implanted in the median nerve only, seven and five EMG electrodes were sutured to the muscles for acute and chronic experiments, respectively. This number would increase if implantation were also performed in the ulnar nerve. The amplifier used for these experiments had eight channels, one of which was used as a synchronisation channel to synchronise the recorded EMGs with the stimulation pattern. Thus, there was no room for additional channels. Even if another amplifier could have been used to record a different set of EMGs, there was not enough time to implant TIMEs in the ulnar nerve, to record several EMGs on the muscles innervated by this nerve, and to follow the stimulation protocol.

4.1.2   Three levels for morphology

For the morphological studies, three levels of the nerves (median and ulnar) were chosen. Because investigation of the complete morphology of the entire length of the nerve starting from the brachial plexus and ending at the finger was not feasible given the timeframe of the project, the region where the electrodes were intended to be implanted was selected. Peripheral nerves begin to branch frequently after the elbow joint and become thinner and flat near the wrist, after which they branch out to innervate the finger muscles. To access most of the muscles in the forearm and hand, a neural electrode should interface with the subpopulation of nerve fibres that innervates these muscles. It was found that for the farm pig, the level just above the elbow joint contained the maximum number of fascicles, and implanting electrodes at that level increased the chance of interfacing with the maximum number of fascicles and thus controlling more muscles. The morphology study performed over the length of the peripheral nerve may not provide complete information about the complete peripheral nerve anatomy of pigs. From human neural anatomy, it can be observed that the peripheral nerve branches out more often just below the elbow joint and innervates various muscles, including those of the fingers. Therefore, at a point just above the elbow joint, the peripheral nerve contains a maximum number of fascicles (a conclusion that has also been found in the farm pig morphology). Therefore, implanting the TIME in this region can provide more selective control over different forearm muscles. The current study provides vital information about the implantation region and the diameters of nerve and fascicles that can be used for designing and optimising neural electrodes in large nerves.

4.1.3   Uses of cuff electrodes

In the present study, a cuff electrode was placed around the median nerve and stimulated with constant current pulses to identify which muscles were activated and to select the optimal locations for the patch electrodes. To optimise the positions of the patch electrodes, these electrodes were repositioned over the muscle surface while the amplitude of the EMG signals was visually monitored with an oscilloscope. After the optimal SNR was determined, the electrode was positioned at that location. This procedure ensured the recording of good EMG signals during stimulation with intrafascicular electrodes. Muscles were included in the experimental selection if they represented different movement characteristics, i.e., flexion, pronation, and abduction of the forelimb, whereas they were excluded if they were small, highly difficult to access, or had very low activations during cuff stimulation.

4.1.4   Stimulation patterns

The stimulation provided in this study was rectangular with a pulse width of 100 μs and frequency of 2 Hz. Monopolar stimulation was performed for acute experiments. The amplitude of the pulse train ranged from 40 μA to 800 μA because the stimulation generator STG2008 was able to produce pulses up to that limit. For chronic experiments, the STG2008 coupled with a DS5 Digitimer stimulator helped to increase the stimulation amplitude to 1,200 μA, which is a safe current injection for TIMEs. Both monopolar and bipolar stimulation configurations were used. Monopolar stimulations resulted in more muscle recruitment at even lower stimulation intensities than bipolar configurations. Bipolar configurations were unable to produce sufficient muscle recruitment in follow-up studies. The histological examination indicated that both the TIME and tfLIFE bypassed the fascicles rather than penetrating them. This situation might cause the failure of bipolar stimulation because the active sites were lying outside the fascicles and were thus unable to recruit muscles above the threshold. Bipolar stimulation produces localised activation, i.e., the current flows in a small, defined, closed path. If the active sites lie outside the fascicles, then it is possible that the current path did not pass through a fascicle and could not activate the muscle. When the stimulation intensity increases, the thickness of the current path travelling from one active site to another also increases. This results in possible interference from the nearest fascicle.

4.1.5   Time window of 2-12 ms

In the acute stimulation selectivity study, the EMG data were analysed by performing the root mean square in a time interval of 2-12 ms after stimulation onset. From previous stimulation studies performed by Kurstjens *et al.* and the authors [85,103], this time window was determined to be ideal for avoiding the influence from stimulation artefacts and for the simultaneous exclusion of any reflex-response during feature extraction and calculation. This time window was narrowed to 3-10 ms for the chronic studies because the EMG recordings from these studies were noisier.

4.1.6   Calculation of the selectivity index

For the acute studies, the selectivity index (SI_M) was calculated as the ratio between the recruitment of one muscle and the sum of the recruitments of all muscles, where the contribution of all muscles was equally weighted. This method has been used in several articles to calculate the SI_M [44,46,104].

$$SI_{M,j} = \frac{EMG(I)_{RL,j}}{\sum_{i=1}^{N} EMG(I)_{RL,i}}$$

However, to calculate selectivity in the chronic studies, a modified version of SI_M was formulated. In the chronic experiments, due to practical issues, the EMG channels were not consistent. Moreover, the SI_M (as calculated in the acute studies) also tends to be more positively biased when there are fewer muscles involved. To prevent the SI_M from increasing due to the lack of EMG response from a channel, a selectivity index for individual muscles with a *constant number* of muscles involved was first defined (SI_{CNM}).

$$SI_{CNM, j} = \frac{EMG(I)_{RL, j}}{EMG(I)_{RL, j} + (N_C - 1) * \left(\dfrac{\sum_{i=1|i\neq j}^{N_A} EMG(I)_{RL, i}}{N_A - 1}\right)}$$

When this constant number N_C was equal to the actual number of muscles N_A available, the selectivity index was calculated in the same manner as for acute studies. As this was not the case, the SI_M for individual muscles in chronic studies was calculated as

$$SI_{M, j} = \frac{N_C}{N_C - 1} * \left(SI(I)_{CNM, j} - \frac{1}{N_C}\right)$$

For both acute and chronic studies, if SI_M is less than zero, then SI_M is equal to zero. The value of $SI_{M,j}$ ranged between zero and one, where zero indicated that muscle j was not activated at all and one indicated that muscle j was the only muscle activated.

4.2 Discussion of the main findings

After performing all four studies, the hypotheses in the introduction were examined along with their major outcomes. The discussions of our findings for each study are summarised below (Table 1).

Table 1: Studies and their major outcomes

Number	Studies	Outcomes
1	Comparison of median and ulnar nerve morphology of Danish landrace pigs and Göttingen mini-pigs	Peripheral nerves of both types of pigs bear closer resemblance to human nerves than other laboratory animals.

2	Comparison of muscle recruitment selectivity during stimulation of the longitudinal or the transverse intrafascicular electrodes (tfLIFE and TIME) in a large nerve animal model	The TIME has a higher selectivity than the tfLIFE with similar activation currents.
3	Evaluation of the chronic stimulation performance of the Transverse, Intrafascicular, Multi-channel Electrode (TIME) in the median nerve of the Göttingen mini-pig	TIMEs were biofunctional and biostable throughout implantation. During chronic experiments, the selectivity decreased and the activation current increased.
4	Assessment of the biocompatibility of the transverse intrafascicular multi-channel electrode (TIME) following sub-chronic implantation in the median nerve of the Göttingen mini-pig	The TIME was found to be biosafe and may be classified as biocompatible for future sub-chronic clinical use.

4.2.1   Morphology

The measurements from the first study provided useful morphological data for designing optimisations to the TIME electrode. This study provided information about the possible nerve regions for implantation of intrafascicular electrodes such that the active sites can interface with the maximum number of fascicles, which will help to control additional muscles. Selecting the right animal model for experiments that bear a closer neuroanatomical resemblance to humans was another objective of this study. Pig forelimb peripheral nerves have morphological similarities with human nerves. Farm pigs were found to be good models for acute experiments. For chronic studies, an animal model that maintains relatively constant physiological and physical features throughout the duration of the study to avoid abnormalities in data analysis is preferred. For this purpose, mini-pigs are better than normal-sized pigs because they are lighter, have an almost constant body weight, do not grow, and are easy to induce to perform simple activities (e.g., walking in a pattern or lying down) if trained.

4.2.2   Selectivity

Selectivity is important for the functional efficiency of neural electrodes. Studies 2 and 3 evaluated the stimulation selectivity of TIMEs in acute and chronic implantations.

During the acute experiments, TIMEs and tfLIFEs were placed in the median nerves of pigs and the recruitment of selected forelimb muscles and the level of activation current required were evaluated. Histological analysis revealed variability in the placement of the electrode inside the nerve and indicated that the electrodes were placed between fascicles rather than within fascicles inside the nerve. Therefore, each active site may activate multiple surrounding fascicles during stimulation, resulting in decreasing selectivity and simultaneously increasing the current required to activate a selected muscle. There was no statistically significant difference in the activation current required by the two electrodes. However, the TIME recruited a greater number of muscles with a higher selectivity than the longitudinally implanted tfLIFE (selectivity index > 0.5).

Chronic experiments are more complex than acute experiments. The chances of obtaining inconsistent results at each follow-up from the same pig are very high. Numerous external and internal factors influence chronic experiments, such as the high impedance of the active sites of the electrode, broken or loose connectors and the behaviour of the pig when infected or when waking up from anaesthesia before the completion of measurements. The average current required to reach the activation threshold in acute experiments was 600 ± 110 µA (mean $\pm$ std), which is above the 488 ± 68 µA value obtained during the first follow-up and even above the I_D of the four TIMEs that remained functional throughout the duration of implantation (552 ± 48 µA). While fibrosis around the implant may not have formed during the first follow-up, the acute I_D was not significantly higher than the I_D in the last follow-up during the chronic experiment.

The SI_D decreased from the first follow-up to the last follow-up, and the number of selectively activated muscles per TIME decreased from 1.17 ± 0.37 to 0.66 ± 0.38. It is not surprising that the selectivity parameters decreased over time due to the formation of a fibrotic capsule around the electrode structure.

The animals supported their weight well on the implanted leg and showed no discomfort following the disappearance of the initial post-operative swelling of the leg. Such visual assessments only rule out severe nerve damage (to motor fibres), as animals might learn to compensate for minor damage. However, these observations were supported by histological analysis, which showed only a localised encapsulation of the polyimide structure of the TIME inside the nerve and no apparent nerve damage.

Differences in the results from our experiments and those of Badia *et al.* were noted [91]. These differences may be due to differences in the nerve models used and/or the methodologies applied. The diameter of the sciatic nerve of the rat (1–1.5 mm) is small and contains only approximately three fascicles. Therefore, the TIME can be inserted directly into the fascicles that innervate the monitored muscles [91,105]. In the multifascicular median nerve of the pig (diameter of 3.5-4 mm, 25-35 fascicles [106]), this setup was not possible. Instead, electrodes were inserted "blindly" into the nerves with no somatotopic knowledge of the fascicle arrangement inside the nerve. The extraneural placement of the TIME has a negative impact on the selectivity parameters in the pig model.

To improve the SI_D, parallel implantation of the TIME in the nerve was later found to be a possible solution [107].

4.2.3   Biocompatibility

In study 4, information about inflammatory cells and fibrosis was used to evaluate the biocompatibility of TIMEs implanted subchronically (approximately five weeks) in Göttingen mini-pigs.

The biofunctionality of TIMEs through chronic experiments showed that selective stimulation gradually decreased over time, and there were no signs of electrode displacement. Formation of a thick fibrosis around the implant could lead to implant failure [108]. The fibrosis thickness at the end of the healing process is typically in the range of 50-200 μm [109]. In this study, all six implantations had fibrotic scarring thicknesses within that range (135 ± 56 μm). Fibrosis thicknesses in this study fell within the normal range. All types of inflammatory cells that typically arise due to foreign body responses were observed. Inflammatory cells were present in large quantities in the implanted region and in small quantities in the control region because the 2 mm distance between the implanted and control regions might look bigger in high powered field but are actually not too large for the real surrounding. Traces of endothelial cells and eosinophils were also found in some samples. Eosinophil cells are phagocytic in nature and develop from host tissue populations or migrate into the tissue from nearby blood vessels [110], while endothelial cells aid in vascular wall remodelling during the healing process of the nerve [111,112]. Endothelial cells are present along the inner surface of the blood vessel. Therefore, the presence of these cells indicates that the TIME pierced through a blood vessel present inside the nerve during its implantation. Overall, the control region of the nerve was normal, with fewer inflammatory cells and without scarring in comparison to the experimental region. Quantitative analysis of fibrosis thickness and semiquantitative analysis of inflammatory cell presence indicate that the TIME is biocompatible and usable for subchronic clinical trials.

4.3 Future work

In all four studies, i.e., morphology of peripheral nerves, stimulation selectivity, and biocompatibility, the median nerve was the nerve of interest. The median nerve is a mixed nerve consisting of both motor and sensory fibres. It innervates several muscles in the forearm, hand, and fingers. The aim of FES rehabilitation is to help patients suffering from paralysis use well-coordinated muscle movements to perform meaningful tasks with their limbs. The upper body limbs perform more complex movements than lower body limbs. Because the median nerve does not innervate all muscles, coordinated movement is not possible. Stimulation via intraneural electrodes should not be applied to the median nerve alone. The ulnar and radial nerves should also be implanted with electrodes simultaneously and selectively stimulated. The whole process requires sophisticated surgery, implantation, and laboratory setups as well as many more studies before clinical trials. As a preliminary experimental setup, TIMEs should be implanted not only in the median but also in the ulnar and radial nerves.

From the selectivity studies and histological sections of the implanted nerves, we found that the TIME does not penetrate the fascicles. This finding directly negatively affects the selectivity and recruitment current of the electrode. No definite solutions have been found, but some possible solutions have been considered, including using a robotic manipulator to shoot the guiding needle of the electrode through the fascicles at a very high speed, which would require a special stereotaxic frame. The region where the TIME is to be implanted would also need to be held steady and tightly (perhaps by a tight cuff electrode) so that the fascicles do not displace when the needle passes through them.

4.4 Conclusions

Improvements to future neural prostheses may be achieved through the availability of multichannel, bidirectional neural interfaces, such as the TIME. This thesis aimed to investigate the overall performance of the TIME in preclinical trials. The biofunctionality, biostability, and biosafety of the TIME were evaluated in the pig model. The results from these studies indicated that the TIME, when implanted in a human-sized nerve, provided higher selectivity than the tfLIFE, was functionally stable during chronic implantation, and caused minimal damage to the nerve, indicating that it is biocompatible with its surrounding tissue.

A balance between the selectivity and invasiveness of neural electrodes is always desirable for the prevention of unwanted outcomes. These results indicate that if the neural interface is biocompatible within its surrounding tissue, it can afford to be invasive to a degree such that it does not harm the physiological and functional properties of the nerve. This increased invasiveness allows the interface to be designed to achieve more selectivity. In general, our results indicate that the TIME has the potential to become a stable interface for use in neuroprosthetic applications for clinical trials.

References

[1]   P. F. Pasquina, P. R. Bryant, M. E. Huang, T. L. Roberts, V. S. Nelson and K. M. Flood, "Advances in amputee care," Arch. Phys. Med. Rehabil., vol. 87, pp. 34-43, 2006.

[2]   P. Parker, K. Englehart and B. Hudgins, "Myoelectric signal processing for control of powered limb prostheses," Journal of Electromyography and Kinesiology, vol. 16, pp. 541-548, 2006.

[3]   F. Andersen-Ranberg and B. Ebskov, "Major upper extremity amputation in Denmark," Acta Orthopaedica, vol. 59, pp. 321-322, 1988.

[4]   M. Martin, "Communication systems for deaf people: a review of possibilities," J. Med. Eng. Technol., vol. 1, pp. 16-28, 1977.

[5]   D. J. Atkins, D. C. Y. Heard and W. H. Donovan, "Epidemiologic overview of individuals with upper-limb loss and their reported research priorities," JPO: Journal of Prosthetics and Orthotics, vol. 8, pp. 2, 1996.

[6]   P. Dario, S. Micera, A. Menciassi, M. Carrozza, M. Zecca, T. Steiglitz, T. Oses, X. Navarro, D. Ceballos and R. Riso, "CYBERHAND-a consortium project for enhanced control of powered artificial hands based on direct neural interfaces," in 33rd Neural Prosthesis Workshop, 2002, .

[7]   X. Navarro, T. B. Krueger, N. Lago, S. Micera, T. Stieglitz and P. Dario, "A critical review of interfaces with the peripheral nervous system for the control of neuroprostheses and hybrid bionic systems," Journal of the Peripheral Nervous System, vol. 10, pp. 229-258, 2005.

[8]   M. P. McLoughlin, "DARPA Revolutionizing Prosthetics 2009," MORS Personnel and National Security Workshop, 2009.

[9]   M. Bridges, J. Beaty, F. Tenore, M. Para, M. Mashner, V. Aggarwal, S. Acharya, G. Singhal and N. Thakor, "Revolutionizing Prosthetics 2009: Dexterous Control of an Upper-Limb Neuroprosthesis," Johns Hopkins APL Tech. Dig., vol. 28, pp. 210, 2010.

[10]   W. M. Grill, "Selective activation of the nervous system for motor system neural prostheses," Intelligent Systems and Technologies in Rehabilitation Engineering, pp. 211-241, 2001.

[11]   N. Nannini and K. Horch, "Muscle recruitment with intrafascicular electrodes," Biomedical Engineering, IEEE Transactions on, vol. 38, pp. 769-776, 1991.

[12]   M. A. Schiefer, R. J. Triolo and D. J. Tyler, "A model of selective activation of the femoral nerve with a flat interface nerve electrode for a lower extremity neuroprosthesis," Neural Systems and Rehabilitation Engineering, IEEE Transactions on, vol. 16, pp. 195-204, 2008.

[13]   M. Schiefer, K. Polasek, R. Triolo, G. Pinault and D. Tyler, "Selective stimulation of the human femoral nerve with a flat interface nerve electrode," Journal of Neural Engineering, vol. 7, pp. 026006, 2010.

38											*References*

[14]	D. Popović and T. Sinkjær, Control of Movement for the Physically Disabled: Control for Rehabilitation Technology. Springer Verlag, 2000.
[15]	D. Rushton, "Functional electrical stimulation," Physiol. Meas., vol. 18, pp. 241, 1997.
[16]	J. Quintern, "Application of functional electrical stimulation in paraplegic patients," NeuroRehabilitation, vol. 10, pp. 205-250, 1998.
[17]	C. L. Lynch and M. R. Popovic, "Functional electrical stimulation," Control Systems, IEEE, vol. 28, pp. 40-50, 2008.
[18]	P. Peckham and G. Creasey, "Neural prostheses: clinical applications of functional electrical stimulation in spinal cord injury," Spinal Cord, vol. 30, pp. 96-101, 1992.
[19]	W. M. Grill and J. T. Mortimer, "Stimulus waveforms for selective neural stimulation," Engineering in Medicine and Biology Magazine, IEEE, vol. 14, pp. 375-385, 1995.
[20]	G. E. Loeb, "Neural prosthetic interfaces with the nervous system," Trends Neurosci., vol. 12, pp. 195-201, 1989.
[21]	P. H. Peckham and J. S. Knutson, "FUNCTIONAL ELECTRICAL STIMULATION FOR NEUROMUSCULAR APPLICATIONS*," Annu. Rev. Biomed. Eng., vol. 7, pp. 327-360, 2005.
[22]	S. F. Cogan, "Neural stimulation and recording electrodes," Annu. Rev. Biomed. Eng., vol. 10, pp. 275-309, 2008.
[23]	M. Popovic, D. Popovic and T. Keller, "Neuroprostheses for grasping," Neurol. Res., vol. 24, pp. 443-452, 2002.
[24]	P. Troyk and N. N. Donaldson, "Implantable fes stimulation systems: What is needed?" Neuromodulation: Technology at the Neural Interface, vol. 4, pp. 196-204, 2001.
[25]	A. Mandal, "What is the nervous system ?" .
[26]	S. S. Sunderland, Nerves and Nerve Injuries. Edinburgh: Churchill Livingstone, 1978.
[27]	W. F. Agnew and D. B. McCreery, "Neural prostheses: fundamental studies," 1990.
[28]	R. Stein, P. Peckham and D. Popovic, "Neural Prostheses," Replacing Motor Function After Disease Or Disability, 1992.
[29]	J. K. Chapin and K. A. Moxon, Neural Prostheses for Restoration of Sensory and Motor Function. CRC, 2000.
[30]	S. Micera, L. Citi, J. Rigosa, J. Carpaneto, S. Raspopovic, G. Di Pino, L. Rossini, K. Yoshida, L. Denaro and P. Dario, "Decoding information from neural signals recorded using intraneural electrodes: Toward the development of a neurocontrolled hand prosthesis," Proc IEEE, vol. 98, pp. 407-417, 2010.
[31]	W. M. Grill, S. E. Norman and R. V. Bellamkonda, "Implanted neural interfaces: biochallenges and engineered solutions," Annu. Rev. Biomed. Eng., vol. 11, pp. 1-24, 2009.
[32]	P. Heiduschka and S. Thanos, "Implantable bioelectronic interfaces for lost nerve functions," Prog. Neurobiol., vol. 55, pp. 433-461, 1998.
[33]	A. Branner, R. B. Stein and R. A. Normann, "Selective stimulation of cat sciatic nerve using an array of varying-length microelectrodes," J. Neurophysiol., vol. 85, pp. 1585-1594, 2001.
[34]	S. Micera, P. Sergi, J. Carpaneto, L. Citi, S. Bossi, K. P. Koch, K. P. Hoffmann, A. Menciassi, K. Yoshida and P. Dario, "Experiments on the development and use of a new generation of intra-neural electrodes to control robotic devices," in Engineering

in Medicine and Biology Society, 2006. EMBS'06. 28th Annual International Conference of the IEEE, 2006, pp. 2940-2943.

[35]   K. P. Hoffmann, K. P. Koch, T. Doerge and S. Micera, "New technologies in manufacturing of different implantable microelectrodes as an interface to the peripheral nervous system," in Biomedical Robotics and Biomechatronics, 2006. BioRob 2006. the First IEEE/RAS-EMBS International Conference on, 2006, pp. 414-419.

[36]   G. Loeb and R. Peck, "Cuff electrodes for chronic stimulation and recording of peripheral nerve activity," J. Neurosci. Methods, vol. 64, pp. 95-103, 1996.

[37]   D. J. Tyler and D. M. Durand, "A slowly penetrating interfascicular nerve electrode for selective activation of peripheral nerves," Rehabilitation Engineering, IEEE Transactions on, vol. 5, pp. 51-61, 1997.

[38]   P. E. Crago, P. H. Peckham and G. B. Thrope, "Modulation of muscle force by recruitment during intramuscular stimulation," Biomedical Engineering, IEEE Transactions on, pp. 679-684, 1980.

[39]   P. A. Grandjean and J. T. Mortimer, "Recruitment properties of monopolar and bipolar epimysial electrodes," Ann. Biomed. Eng., vol. 14, pp. 53-66, 1986.

[40]   T. Boretius, J. Badia, A. Pascual-Font, M. Schuettler, X. Navarro, K. Yoshida and T. Stieglitz, "A transverse intrafascicular multichannel electrode (TIME) to interface with the peripheral nerve," Biosensors and Bioelectronics, vol. 26, pp. 62-69, 2010.

[41]   C. Gonzalez and M. Rodriguez, "A flexible perforated microelectrode array probe for action potential recording in nerve and muscle tissues," J. Neurosci. Methods, vol. 72, pp. 189-195, 1997.

[42]   W. Grill and J. T. Mortimer, "Quantification of recruitment properties of multiple contact cuff electrodes," Rehabilitation Engineering, IEEE Transactions on, vol. 4, pp. 49-62, 1996.

[43]   W. M. Grill and J. T. Mortimer, "Stability of the input-output properties of chronically implanted multiple contact nerve cuff stimulating electrodes," Rehabilitation Engineering, IEEE Transactions on, vol. 6, pp. 364-373, 1998.

[44]   T. Nielsen, M. Kurstjens and J. Struijk, "Transverse vs. Longitudinal Tripolar Configuration for Selective Stimulation with Multipolar Cuff Electrodes," Biomedical Engineering, IEEE Transactions on, vol. 58, pp. 913-919, 2011.

[45]   M. Kurstjens and W. Jensen, "Selectivity of longitudinal versus transverse tripolar stimulation of median nerve in pigs using a multicontact nerve cuff electrode," .

[46]   C. Veraart, W. M. Grill and J. T. Mortimer, "Selective control of muscle activation with a multipolar nerve cuff electrode," Biomedical Engineering, IEEE Transactions on, vol. 40, pp. 640-653, 1993.

[47]   K. H. Polasek, H. A. Hoyen, M. W. Keith, R. F. Kirsch and D. J. Tyler, "Stimulation stability and selectivity of chronically implanted multicontact nerve cuff electrodes in the human upper extremity," Neural Systems and Rehabilitation Engineering, IEEE Transactions on, vol. 17, pp. 428-437, 2009.

[48]   K. H. Polasek, H. A. Hoyen, M. W. Keith and D. J. Tyler, "Human nerve stimulation thresholds and selectivity using a multi-contact nerve cuff electrode," Neural Systems and Rehabilitation Engineering, IEEE Transactions on, vol. 15, pp. 76-82, 2007.

[49]   J. A. Hoffer, R. B. Stein, M. K. Haugland, T. Sinkjaer, W. K. Durfee, A. B. Schwartz, G. E. Loeb and C. Kantor, "Neural signals for command control and

feedback in functional neuromuscular stimulation: a review," J. Rehabil. Res. Dev., vol. 33, pp. 145-157, 1996.

[50] J. Rozman, R. Acimovic-Janezic, I. Tekavcic, M. Kljajic and M. Trlep, "Implantable stimulator for selective stimulation of the common peroneal nerve: a preliminary report," J. Med. Eng. Technol., vol. 18, pp. 47-53, 1994.

[51] P. J. Slot, P. Selmar, A. Rasmussen and T. Sinkjaer, "Effect of long-term implanted nerve cuff electrodes on the electrophysiological properties of human sensory nerves," Artif. Organs, vol. 21, pp. 207-209, Mar, 1997.

[52] M. Rahal, J. Taylor and N. Donaldson, "The effect of nerve cuff geometry on interference reduction: a study by computer modeling," Biomedical Engineering, IEEE Transactions on, vol. 47, pp. 136-138, 2000.

[53] A. Branner, R. B. Stein, E. Fernandez, Y. Aoyagi and R. A. Normann, "Long-term stimulation and recording with a penetrating microelectrode array in cat sciatic nerve," Biomedical Engineering, IEEE Transactions on, vol. 51, pp. 146-157, 2004.

[54] W. M. Grill and J. T. Mortimer, "Neural and connective tissue response to long - term implantation of multiple contact nerve cuff electrodes," J. Biomed. Mater. Res., vol. 50, pp. 215-226, 2000.

[55] D. Ceballos, A. Valero, E. Valderrama, T. Stieglitz and X. Navarro, "Polyimide cuff electrodes for peripheral nerve stimulation," J. Neurosci. Methods, vol. 98, pp. 105-118, 2000.

[56] H. Thoma, W. Girsch, J. Holle and W. Mayr, "Technology and long-term application of an epineural electrode." ASAIO transactions/American Society for Artificial Internal Organs, vol. 35, pp. 490, 1989.

[57] D. J. Tyler and D. M. Durand, "Functionally selective peripheral nerve stimulation with a flat interface nerve electrode," Neural Systems and Rehabilitation Engineering, IEEE Transactions on, vol. 10, pp. 294-303, 2002.

[58] D. K. Leventhal and D. M. Durand, "Subfascicle stimulation selectivity with the flat interface nerve electrode," Ann. Biomed. Eng., vol. 31, pp. 643-652, 2003.

[59] D. K. Leventhal and D. M. Durand, "Chronic measurement of the stimulation selectivity of the flat interface nerve electrode," Biomedical Engineering, IEEE Transactions on, vol. 51, pp. 1649-1658, 2004.

[60] D. K. Leventhal, M. Cohen and D. M. Durand, "Chronic histological effects of the flat interface nerve electrode," Journal of Neural Engineering, vol. 3, pp. 102, 2006.

[61] D. J. Tyler and D. M. Durand, "Chronic response of the rat sciatic nerve to the flat interface nerve electrode," Ann. Biomed. Eng., vol. 31, pp. 633-642, 2003.

[62] H. Powell and R. Myers, "Pathology of experimental nerve compression." Lab. Invest., vol. 55, pp. 91, 1986.

[63] D. J. Tyler and D. M. Durand, "Interfascicular electrical stimulation for selectively activating axons," IEEE Eng. Med. Biol., vol. 13, pp. 575-583, 1994.

[64] D. J. Tyler and D. Durand, "Design and acute test of a radially penetrating interfascicular nerve electrode," in Engineering in Medicine and Biology Society, 1993. Proceedings of the 15th Annual International Conference of the IEEE, 1993, pp. 1247-1248.

[65] K. Yoshida and K. Horch, "Selective stimulation of peripheral nerve fibers using dual intrafascicular electrodes," Biomedical Engineering, IEEE Transactions on, vol. 40, pp. 492-494, 1993.

[66] K. Yoshida, K. Jovanovic and R. B. Stein, "Intrafascicular electrodes for stimulation and recording from mudpuppy spinal roots," J. Neurosci. Methods, vol. 96, pp. 47-55, 2000.

[67] K. Yoshida, D. Pellinen, D. Pivin, P. Rousche and D. Kipke, "Development of the thin-film longitudinal intra-fascicular electrode," in Proceedings of the Fifth Annual Conf. of the IFESS, 2000, pp. 279-284.

[68] W. Jensen, S. Micera, X. Navarro, T. Stieglitz, D. Guiraud, J. L. Divoux, P. Rossini and K. Yoshida, "S19. 5 transverse, intrafascicular multi-electrode (TIME) system for induction of sensation and treatment of phantom limb pain in amputees," in 2011, pp. S46-S46.

[69] A. Branner and R. A. Normann, "A multielectrode array for intrafascicular recording and stimulation in sciatic nerve of cats," Brain Res. Bull., vol. 51, pp. 293-306, 2000.

[70] S. Micera, X. Navarro, J. Carpaneto, L. Citi, O. Tonet, P. M. Rossini, M. C. Carrozza, K. P. Hoffmann, M. Vivó and K. Yoshida, "On the use of longitudinal intrafascicular peripheral interfaces for the control of cybernetic hand prostheses in amputees," Neural Systems and Rehabilitation Engineering, IEEE Transactions on, vol. 16, pp. 453-472, 2008.

[71] R. A. Freitas Jr, "Volume IIA: Biocompatibility," 2003.

[72] W. Jensen and K. Yoshida, "Long-term recording properties of longitudinal intra-fascicular electrodes," in 7th Annual Conference of the International Functional Electrical Stimulation Society, IFESS, Ljubljana, Slovenia, 25-29 June 2002, pp. 138-140.

[73] T. G. McNaughton and K. W. Horch, "A dual-channel intrafascicular electrode for tracking single cell activities in peripheral nerves," in Engineering in Medicine and Biology Society, 1992 14th Annual International Conference of the IEEE, 1992, pp. 1326-1327.

[74] M. S. Malagodi, K. W. Horch and A. A. Schoenberg, "An intrafascicular electrode for recording of action potentials in peripheral nerves," Ann. Biomed. Eng., vol. 17, pp. 397-410, 1989.

[75] T. McNaughton and K. Horch, "Action potential classification with dual channel intrafascicular electrodes," Biomedical Engineering, IEEE Transactions on, vol. 41, pp. 609-616, 1994.

[76] T. Lefurge, E. Goodall, K. Horch, L. Stensaas and A. Schoenberg, "Chronically implanted intrafascicular recording electrodes," Ann. Biomed. Eng., vol. 19, pp. 197-207, 1991.

[77] K. Yoshida and K. Horch, "Closed-loop control of ankle position using muscle afferent feedback with functional neuromuscular stimulation," Biomedical Engineering, IEEE Transactions on, vol. 43, pp. 167-176, 1996.

[78] G. Dhillon, T. Krüger, J. Sandhu and K. Horch, "Effects of short-term training on sensory and motor function in severed nerves of long-term human amputees," J. Neurophysiol., vol. 93, pp. 2625-2633, 2005.

[79] K. Horch, S. Meek, T. G. Taylor and D. Hutchinson, "Object discrimination with an artificial hand using electrical stimulation of peripheral tactile and proprioceptive pathways with intrafascicular electrodes," Neural Systems and Rehabilitation Engineering, IEEE Transactions on, vol. 19, pp. 483-489, 2011.

[80]   G. S. Dhillon, S. M. Lawrence, D. T. Hutchinson and K. W. Horch, "Residual function in peripheral nerve stumps of amputees: implications for neural control of artificial limbs1," J. Hand Surg., vol. 29, pp. 605-615, 2004.

[81]   G. S. Dhillon and K. W. Horch, "Direct neural sensory feedback and control of a prosthetic arm," Neural Systems and Rehabilitation Engineering, IEEE Transactions on, vol. 13, pp. 468-472, 2005.

[82]   N. Lago, K. Yoshida, K. P. Koch and X. Navarro, "Assessment of biocompatibility of chronically implanted polyimide and platinum intrafascicular electrodes," Biomedical Engineering, IEEE Transactions on, vol. 54, pp. 281-290, 2007.

[83]   L. Citi, J. Carpaneto, K. Yoshida, K. Hoffmann, K. Koch, P. Dario and S. Micera, "Characterization of tfLIFE neural response for the control of a cybernetic hand," in Biomedical Robotics and Biomechatronics, 2006. BioRob 2006. the First IEEE/RAS-EMBS International Conference on, 2006, pp. 477-482.

[84]   K. Yoshida, M. Kurstjens, L. Citi, K. P. Koch and S. Micera, "Recording experience with the thin-film longitudinal intra-fascicular electrode, a multichannel peripheral nerve interface," in Rehabilitation Robotics, 2007. ICORR 2007. IEEE 10th International Conference on, 2007, pp. 862-867.

[85]   M. Kurstjens, W. Jensen and K. Yoshida, "Selective activation of pig forearm muscles using thin-film intrafascicular electrodes implanted in the median nerve," in 13th International FES Society Conference, 2008, pp. 233-235.

[86]   X. Navarro, N. Lago, M. Vivo, K. Yoshida, K. P. Koch, W. Poppendieck and S. Micera, "Neurobiological evaluation of thin-film longitudinal intrafascicular electrodes as a peripheral nerve interface," in Rehabilitation Robotics, 2007. ICORR 2007. IEEE 10th International Conference on, 2007, pp. 643-649.

[87]   A. Benvenuto, S. Raspopovic, K. Hoffmann, J. Carpaneto, G. Cavallo, G. Di Pino, E. Guglielmelli, L. Rossini, P. Rossini and M. Tombini, "Intrafascicular thin-film multichannel electrodes for sensory feedback: Evidences on a human amputee," in Engineering in Medicine and Biology Society (EMBC), 2010 Annual International Conference of the IEEE, 2010, pp. 1800-1803.

[88]   P. M. Rossini, S. Micera, A. Benvenuto, J. Carpaneto, G. Cavallo, L. Citi, C. Cipriani, L. Denaro, V. Denaro and G. Di Pino, "Double nerve intraneural interface implant on a human amputee for robotic hand control," Clinical Neurophysiology, vol. 121, pp. 777-783, 2010.

[89]   A. Kundu, K. R. Harreby, M. Kurstjens, T. Boretius, T. Stieglitz, K. Yoshida and W. Jensen, "Comparison of acute stimulation selectivity of transverse and longitudinal intrafascicular electrodes in pigs," in 2011, .

[90]   W. Jensen, S. Micera, X. Navarro, T. Stieglitz, D. Guiraud, J. Divoux, P. Rossini and K. Yoshida, "Development of an implantable transverse intrafascicular multichannel electrode (TIME) system for relieving phantom limb pain," in Engineering in Medicine and Biology Society (EMBC), 2010 Annual International Conference of the IEEE, 2010, pp. 6214-6217.

[91]   J. Badia, T. Boretius, D. Andreu, C. Azevedo-Coste, T. Stieglitz and X. Navarro, "Comparative analysis of transverse intrafascicular multichannel, longitudinal intrafascicular and multipolar cuff electrodes for the selective stimulation of nerve fascicles," Journal of Neural Engineering, vol. 8, pp. 036023, 2011.

[92]   T. Boretius, J. Badia, A. Pascual-Font, D. Andreu, C. Azevedo-Coste, D. Jean-Louis, T. Stieglitz, X. Navarro and K. Yoshida, "Transverse intrafascicular multichannel electrode (TIME) an interface to peripheral nerves: Preliminary in-

vivo results in rats," in Proceedings of the 14th IFESS Conference, Seoul, South Korea, 2009, pp. 37-39.

[93]   T. Boretius, K. Yoshida, J. Badia, K. Harreby, A. Kundu, X. Navarro, W. Jensen and T. Stieglitz, "A transverse intrafascicular multichannel electrode (TIME) to treat phantom limb pain – towards human clinical trials," in IEEE International Conference on Biomedical Robotics and Biomechatronics, June 24-28, Roma, Italy, 2012, pp. 282-287.

[94]   J. B. Phillips, X. Smit, N. D. Zoysa, A. Afoke and R. A. Brown, "Peripheral nerves in the rat exhibit localized heterogeneity of tensile properties during limb movement," J. Physiol. (Lond.), vol. 557, pp. 879-887, 2004.

[95]   F. Panetsos, C. Avendaño, P. Negredo, J. Castro and V. Bonacasa, "Neural prostheses: electrophysiological and histological evaluation of central nervous system alterations due to long-term implants of sieve electrodes to peripheral nerves in cats," Neural Systems and Rehabilitation Engineering, IEEE Transactions on, vol. 16, pp. 223-232, 2008.

[96]   S. E. Mackinnon and A. Lee Dellon, "Evaluation of microsurgical internal neurolysis in a primate median nerve model of chronic nerve compression," J. Hand Surg., vol. 13, pp. 345-351, 1988.

[97]   P. M. Fullerton and R. Gilliatt, "Median and ulnar neuropathy in the guinea-pig." Journal of Neurology, Neurosurgery & Psychiatry, vol. 30, pp. 393-402, 1967.

[98]   J. Forden, Q. G. Xu, K. J. Khu and R. Midha, "A long peripheral nerve autograft model in the sheep forelimb," Neurosurgery, vol. 68, pp. 1354, 2011.

[99]   S. F. Cogan, Y. Liu and J. D. Walter, "THIN-FILM PERIPHERAL NERVE ELECTRODE:Fabrication of nerve cuffs and evaluation of selective stimulationand somatosensory neuroprostheses in a raccoon model," 1997.

[100] Staff of ADInstruments, "Action potentials in earthworm giant nerve fibers," ADInstruments, .

[101] J. Badia, T. Boretius, A. Pascual-Font, E. Udina, T. Stieglitz and X. Navarro, "Biocompatibility of chronically implanted transverse intrafascicular multichannel electrode (TIME) in the rat sciatic nerve," Biomedical Engineering, IEEE Transactions on, vol. 58, pp. 2324-2332, 2011.

[102] R. Riso, A. Dalmose, M. Schüttler and T. Stieglitz, "Activation of muscles in the pig forelimb using a large diameter multipolar nerve cuff installed on the radial nerve in the axilla," in Proc. V Intern IFESS Conf, Aalborg, 2000, pp. 272-275.

[103] A. Kundu, W. Jensen, M. Kurstjens, T. Stieglitz, T. Boretius and K. Yoshida, "Dependence of implantation angle of the transverse, intrafascicular electrode (TIME) on selective activation of pig forelimb muscles," Artif. Organs, vol. 34, pp. A43, 2010.

[104] P. B. Yoo, M. Sahin and D. M. Durand, "Selective stimulation of the canine hypoglossal nerve using a multi-contact cuff electrode," Ann. Biomed. Eng., vol. 32, pp. 511-519, 2004.

[105] J. Badia, A. Pascual ‐ Font, M. Vivó, E. Udina and X. Navarro, "Topographical distribution of motor fascicles in the sciatic ‐ tibial nerve of the rat," Muscle Nerve, vol. 42, pp. 192-201, 2010.

[106] Kundu A., K. R. Harreby and W. Jensen, "Comparison of median and ulnar nerve mrorphology of danish landrace pigs and göttingen mini pigs," in 17th IFESS Annual Conference, 9-12 September, Banff, Alberta, Canada, 2012, .

[107]   K. R. Harreby, A. Kundu, B. Geng, P. Maciejasz, D. Guiraud, T. Stieglitz, T. Boretius, K. Yoshida and W. Jensen, "Recruitment selectivity of single and pairs of transverse, intrafascicular, multi-channel electrodes (TIME) in the pig median nerve," in IFESS'2012: 17th International Conference of Functional Electrical Stimulation Society, 2012, .

[108]   C. Jensen, L. Gurevich, A. Patriciu, J. Struijk, V. Zachar and C. P. Pennisi, "Increased connective tissue attachment to silicone implants by a water vapor plasma treatment," Journal of Biomedical Materials Research Part A, 2012.

[109]   J. M. Morais, F. Papadimitrakopoulos and D. J. Burgess, "Biomaterials/tissue interactions: possible solutions to overcome foreign body response," The AAPS Journal, vol. 12, pp. 188-196, 2010.

[110]   B. Young, W. Stewart and G. O'Dowd, Wheater's Basic Pathology: A Text, Atlas and Review of Histopathology. Churchill Livingstone, 2011.

[111]   C. Michiels, "Endothelial cell functions," J. Cell. Physiol., vol. 196, pp. 430-443, 2003.

[112]   K. Peters, R. E. Unger, C. J. Kirkpatrick, A. M. Gatti and E. Monari, "Effects of nano-scaled particles on endothelial cell function in vitro: studies on viability, proliferation and inflammation," J. Mater. Sci. Mater. Med., vol. 15, pp. 321-325, 2004.

www.ingramcontent.com/pod-product-compliance
Lightning Source LLC
Chambersburg PA
CBHW061840220326
41599CB00027B/5359